Written by Megan Noel
Cover design by Tina DeKam


goodandbeautiful.com

Looking up into a clear night sky, you can sometimes catch a glimpse of a red-tinted sphere among the twinkling stars. That mysterious object is the fourth planet from the sun: Mars. If you have ever wondered what it might be like up there on the appropriately nicknamed Red Planet, you are not alone.

Since 1960, many scientists have made attempts to explore one of our planet's nearest neighbors. Many of these attempts were unsuccessful because of problems with launching, equipment failures, loss of contact, and other issues. However, scientific pioneers refused to give up. The determination of these great minds has led to many amazing discoveries and a much greater understanding of this fascinating planet than we could have ever before anticipated.

The Soviet Union (USSR), was the first to try to send a spacecraft to Mars, but Marsnik 1 failed to reach Earth's orbit. After several more unsuccessful tries by both the USSR and the United States, the USA sent Mariner 4 on the first successful flyby mission.

In 1965, after completing an eight-month journey, Mariner 4 was able to come within 9,844 km (6,118 mi) of Mars' cratered surface. Mariner 4 sent us our very first close-up pictures of Mars' surface as well as information about its thin atmosphere, which was confirmed to be composed of carbon dioxide.

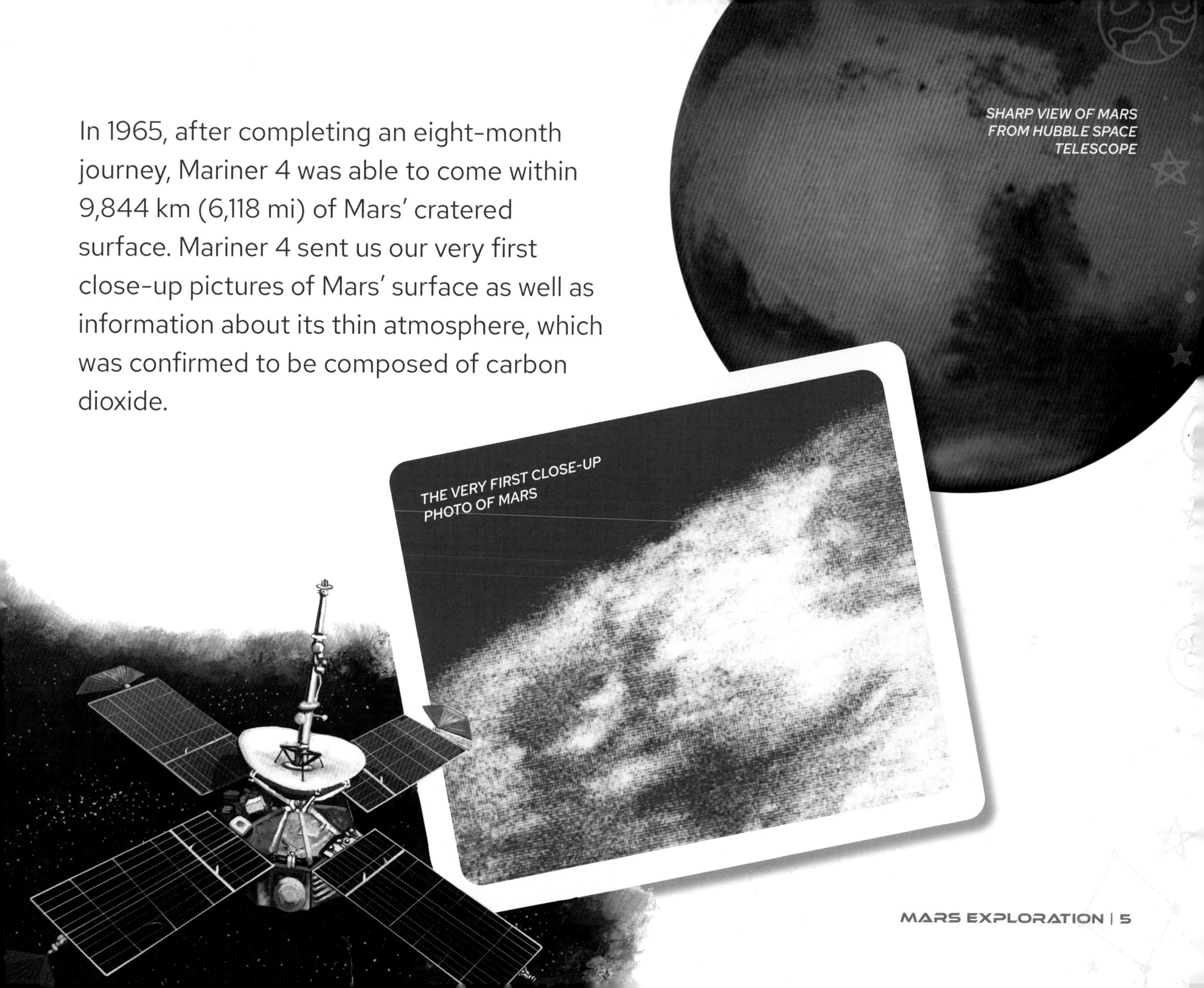

SHARP VIEW OF MARS FROM HUBBLE SPACE TELESCOPE

THE VERY FIRST CLOSE-UP PHOTO OF MARS

The Mariner missions continued in 1969 when the USA sent the Mariner 6 and Mariner 7 on the first dual mission to Mars to take measurements of the temperature, pressure, and composition of Mars' atmosphere. The world gained 200 new pictures of Mars from this mission, which gave us views of the equator and south polar regions.

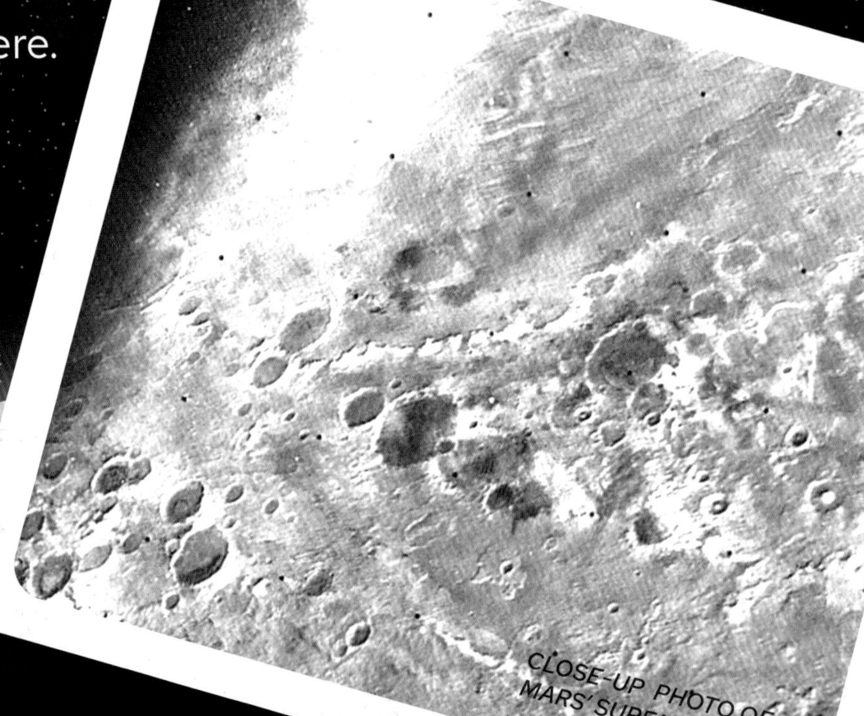

CLOSE-UP PHOTO OF MARS' SURFACE

In 1971 the USSR and USA both achieved firsts in Mars exploration. The Soviet probe, Mars 3, made the first successful landing on Mars, although it failed after relaying just 20 seconds of video back to Earth. The first spacecraft to orbit another planet was the USA's Mariner 9.

THE MARINER 9 LAUNCH TOWARD MARS FROM CAPE KENNEDY'S LAUNCH COMPLEX ON MAY 30, 1971

THE MARINER 9

After a month-long dust storm finally cleared, Mariner 9 was able to send back a sizable collection of high-quality photographs that taught us many new things about the dry and dusty surface of Mars. We were able to see towering volcanoes, a vast canyon stretching 4,828 km (3,000 mi), and ancient river beds through which liquid water likely once flowed.

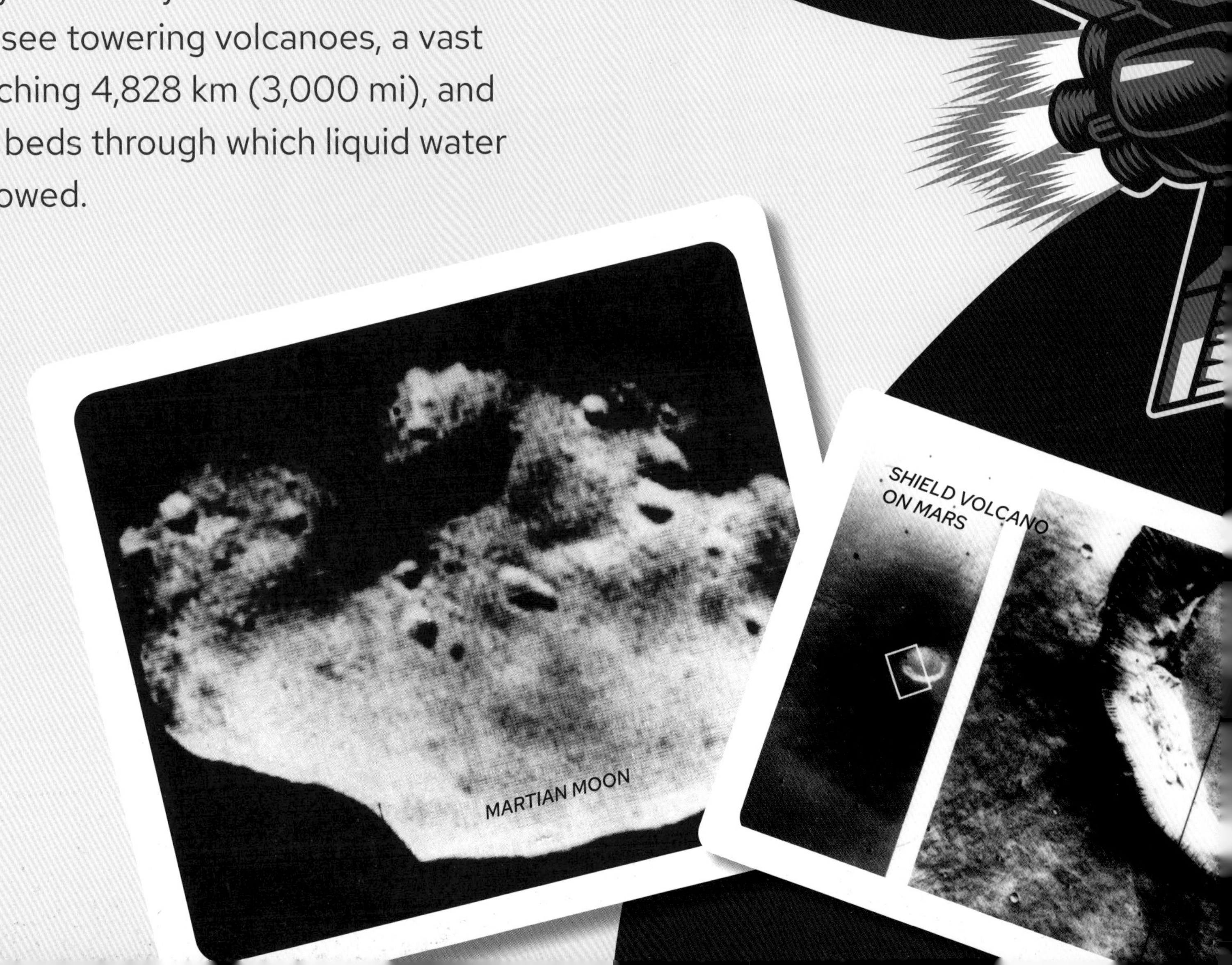

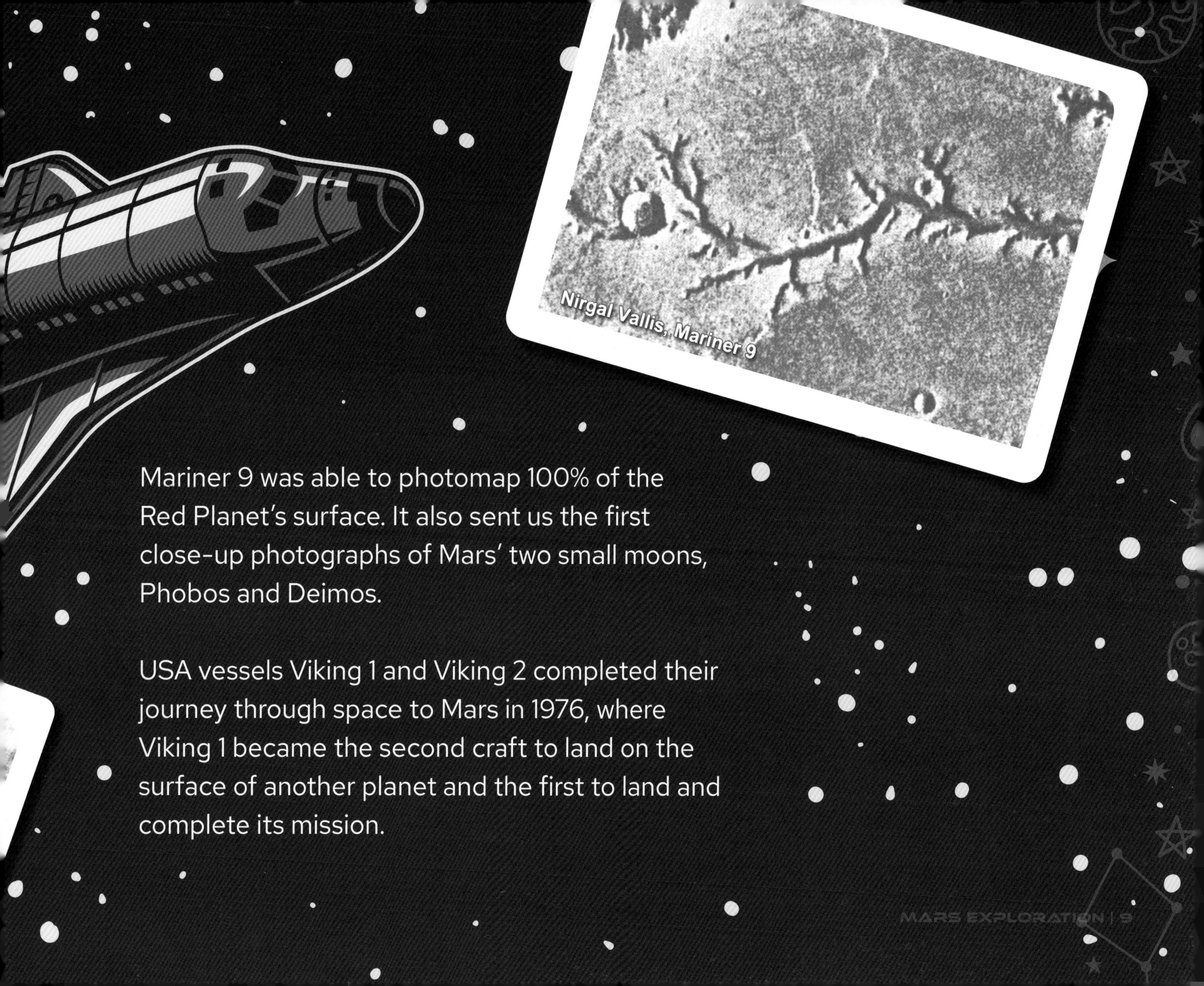

Nirgal Vallis, Mariner 9

Mariner 9 was able to photomap 100% of the Red Planet's surface. It also sent us the first close-up photographs of Mars' two small moons, Phobos and Deimos.

USA vessels Viking 1 and Viking 2 completed their journey through space to Mars in 1976, where Viking 1 became the second craft to land on the surface of another planet and the first to land and complete its mission.

*PANORAMIC VIEW FROM THE VIKING 1*

Both Viking spacecrafts contained an orbiter and a lander. The orbiters relayed an impressive 14,000+ images. The landers monitored the weather and performed experiments in an effort to seek out living microorganisms. The results were not entirely clear, but it is now widely believed that Mars, as it is today, is unable to support life. Still, that hasn't stopped other missions from continuing the search for evidence of life in its past.

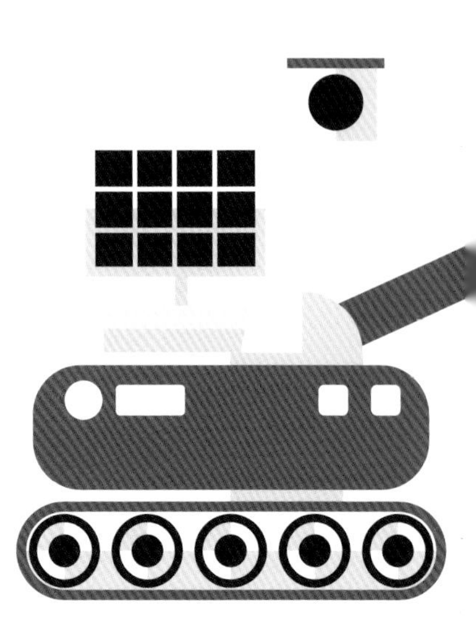

The Mars Pathfinder mission of the mid-1990s was part of NASA's Discovery series, whose primary purpose was to show that low-cost landings on Mars were possible. Mars Pathfinder's stationary lander and NASA's first Mars rover, Sojourner, were successful in proving this, and this mission paved the way for later Discovery missions such as the more recent InSight. As we continue to explore our vast universe, cost-effectiveness will surely be an important consideration.

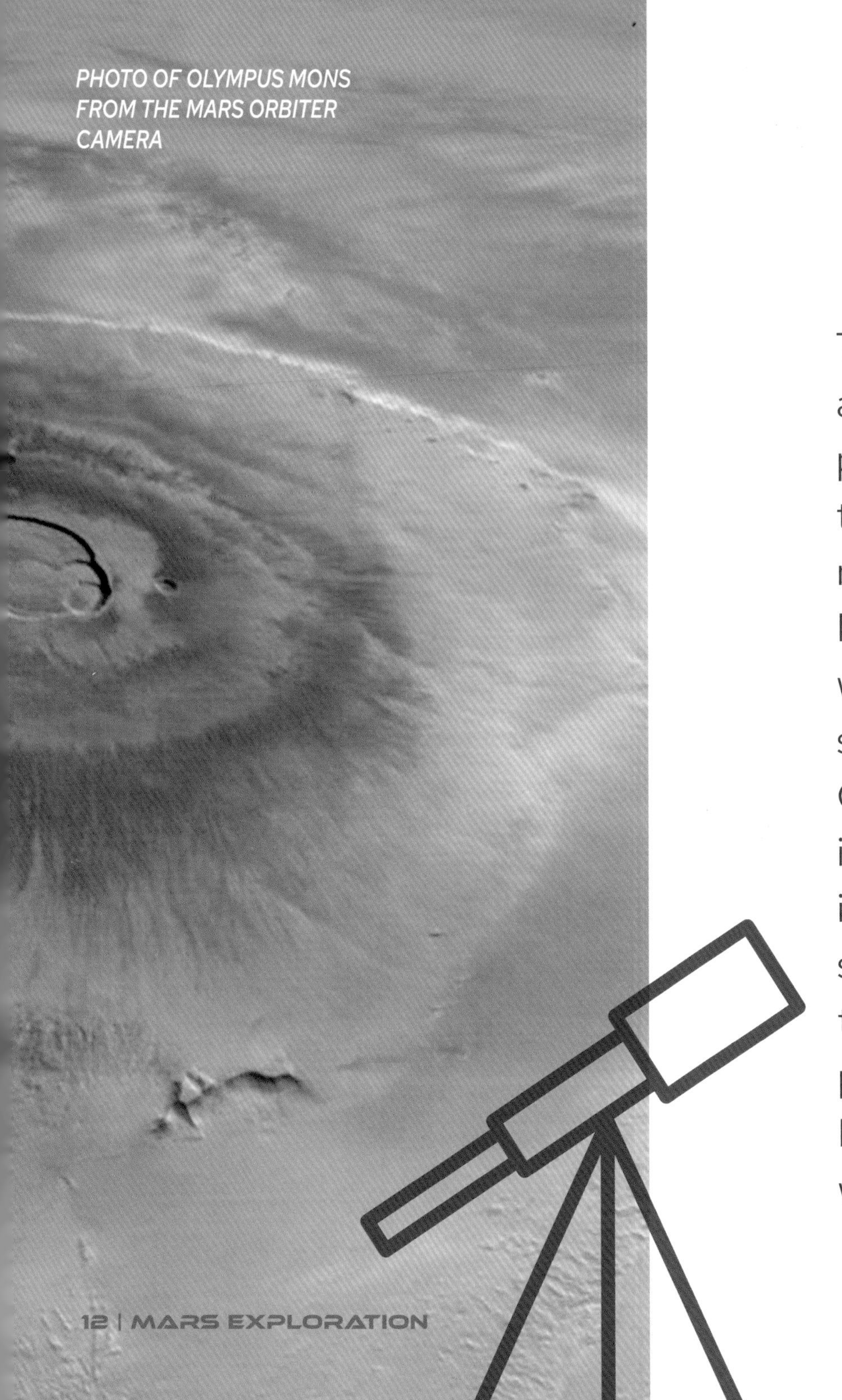
*PHOTO OF OLYMPUS MONS FROM THE MARS ORBITER CAMERA*

The Mars Global Surveyor began its mission as a Martian meteorologist in 1996, studying the planet's frigid weather. One of its instruments took measurements of Olympus Mons, which not only proved to be the largest volcano on Mars but the largest known volcano in our whole solar system! Though its gradually sloping sides make it appear quite flat, Olympus Mons is absolutely massive. It is about as wide as the entire Hawaiian island chain (2,414 km/1,500 mi) and stands 28 km (16 mi) high, which is taller than three Mount Everests. This is possible because of Mars' low gravity. In Earth's gravity, a mountain that size would collapse under its own weight.

The USA landed solar-powered identical twin rovers, Spirit and Opportunity, on Mars in 2004. Their primary purpose was to search for more evidence of the liquid water—and potentially life—that scientists believe existed there a long time ago. Communication with Spirit was lost in 2010.

MARS EXPLORATION ROVER

But Opportunity was a record breaker, spending about 14 and a half years exploring Mars' surface. This feat was quite a surprise to everyone since the rover was only designed to operate for 90 days. It was also only expected to travel 1 km (1,100 yd), yet it went on to explore 45 km (28 mi) of the arid landscape. It is largely because of Opportunity's many accomplishments that we may someday be able to send astronauts to study Mars firsthand.

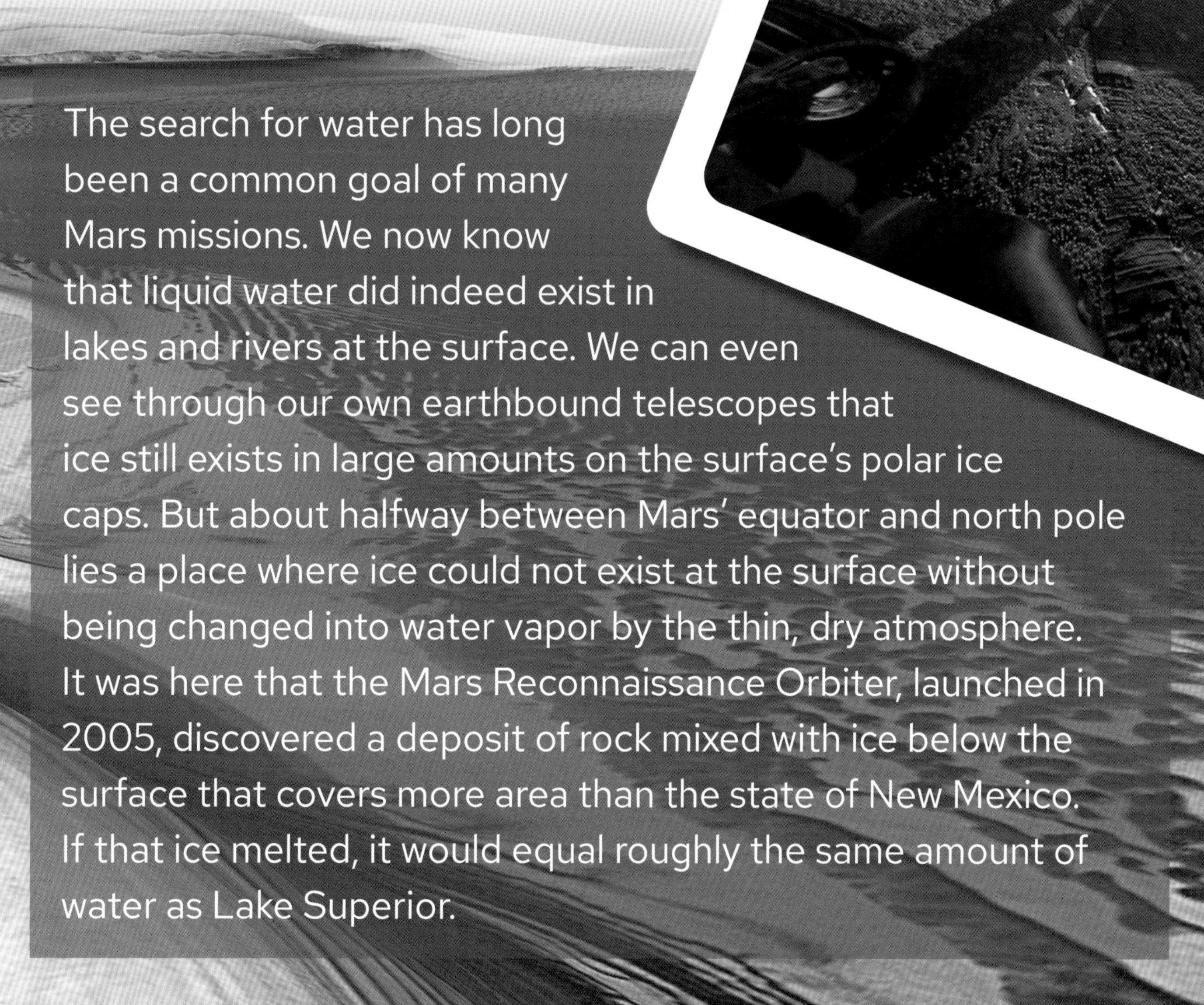

OPPORTUNITY'S SHADOW ON MARS

The search for water has long been a common goal of many Mars missions. We now know that liquid water did indeed exist in lakes and rivers at the surface. We can even see through our own earthbound telescopes that ice still exists in large amounts on the surface's polar ice caps. But about halfway between Mars' equator and north pole lies a place where ice could not exist at the surface without being changed into water vapor by the thin, dry atmosphere. It was here that the Mars Reconnaissance Orbiter, launched in 2005, discovered a deposit of rock mixed with ice below the surface that covers more area than the state of New Mexico. If that ice melted, it would equal roughly the same amount of water as Lake Superior.

*COMPUTER SIMULATION OF POLAR ICE CAPS*

In 2012, NASA's Mars Science Laboratory continued the detective work of searching for clues as to whether life once thrived on Mars. The Mars Science Laboratory was special because it used a brand-new method to land and release its rover, Curiosity.

*AN ARTIST'S RENDERING OF CURIOSITY LANDING ON THE MOON.*

The laboratory descended to the surface on a parachute, then seconds before it landed, rockets were fired which allowed it to hover long enough to safely lower Curiosity to the ground by way of a tether. After Curiosity's successful landing, the Mars Science Laboratory, having fulfilled its duty, flew away for a crash landing.

THE CURIOSITY ROVER'S ENTRY, DESCENT, AND LANDING TEAM (EDL) ARRIVES AT THE POST-LANDING PRESS CONFERENCE TO CHEERS AND CELEBRATION.

Once Curiosity started rolling, it was able to use its scientific tools to discover chemical and mineral evidence that Mars could have indeed been habitable long ago. It has drilled into Mars' rock layers for many years, seeking signs of microbial life. Curiosity was also the first rover to take a selfie on another planet!

A CLOSE-UP OF CURIOSITY

MARS SELFIE

Would you believe that a spacecraft has been sent to Mars with the hope of gaining a better understanding of Earth's atmosphere? It's true! If you've ever been listening to the radio and heard static interference that sounded like you were hearing another station over your own music, your radio's signal may have been interrupted by electrically charged plasma layers in our ionosphere. This may also be related to the aurora borealis ("Northern Lights") which also occur in the ionosphere. Exploration of Earth's ionosphere would be extremely difficult here because it is too thin for airplane travel, yet it is too thick for a satellite to orbit without burning up.

*AURORA BOREALIS*

So why would we send a spacecraft about 225 million km (140 million mi) away to study that very same phenomenon? Mars' atmosphere is thinner than our own, making it much easier to explore.

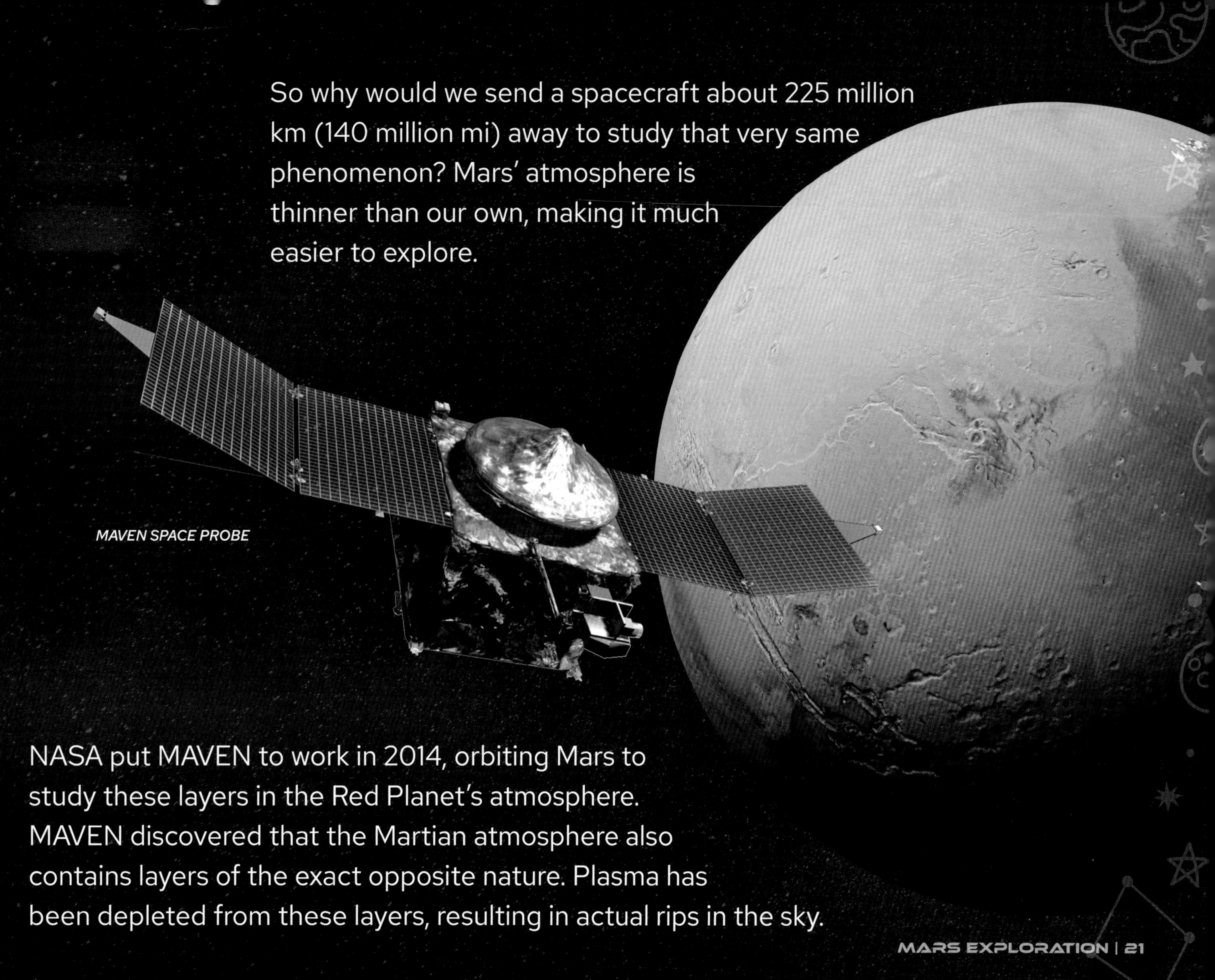

*MAVEN SPACE PROBE*

NASA put MAVEN to work in 2014, orbiting Mars to study these layers in the Red Planet's atmosphere. MAVEN discovered that the Martian atmosphere also contains layers of the exact opposite nature. Plasma has been depleted from these layers, resulting in actual rips in the sky.

MAVEN was also tasked with teaching us more about the drastic climate change Mars has experienced over time. By mapping wind speeds high in Mars' atmosphere, we know more about how the features of Mars were formed, including its many volcanoes. There is still so much to learn about the ancient history of Mars and why it is no longer a warm, wet world that could have likely supported life. MAVEN is helping scientists connect the pieces of this puzzle.

Although the United States and Soviet Union/Russia have sent up the vast majority of spacecraft with which to study Mars up close, they are not the only countries that have tried. Japan's Nozomi and China's Yinghuo-1 orbiters attempted to reach Mars, but sadly neither one was able to carry out its missions. However, Nozomi did provide us with more information about the environment in deep space. India did succeed in inserting the Mars Orbiter Mission (also known as Mangalyaan) into Mars' orbit in 2014. The orbiter was successful in reaching its goal of mapping the planet and measuring radiation.

Mars exploration continues to this day, and it shows no signs of coming to an end. Many spacecraft and rovers have provided us with invaluable information that, rather than satisfying our curiosity, only fuels it all the more. For every question answered, there are always more questions being asked and more answers to be found.

Perhaps one day we will land the first brave human expedition on Mars. Until then, we will continue on our quest for more knowledge about this strange Martian world.